SpringerBriefs in Energy

More information about this series at http://www.springer.com/series/8903

Uttam Roy · Mrinmoy Majumder

Impact of Climate Change on Small Scale Hydro-turbine Selections

 Springer

Uttam Roy
Bengal Institute of Technology
 and Management
Kolkata
India

Mrinmoy Majumder
National Institute of Technology
Agartala
India

ISSN 2191-5520   ISSN 2191-5539   (electronic)
SpringerBriefs in Energy
ISBN 978-981-287-238-8   ISBN 978-981-287-239-5   (eBook)
DOI 10.1007/978-981-287-239-5

Library of Congress Control Number: 2016936979

Printed on acid-free paper

This Springer imprint is published by Springer Nature
The registered company is Springer Science+Business Media Singapore Pte Ltd.

Preface

The efficiency of turbines which are used in hydropower plants also ensures the overall production efficiency of hydropower plants. The abnormalities in climatic pattern which is growing day by day can also impact the performance of hydro-turbines (Majone et al. 2016). The hydropower plants depend on flowing water and the runners of such kind of plant is either submerged or get into touch of water. The quantity (Wang et al. 2016) as well as quality (Kumar and Sarkar 2016) of water depends on climatic parameters like rainfall and temperature. Both quantity and quality of water will also affect the performance of the turbine. Frictions and corrosion caused by the flowing water on turbines will increase as the quantity and quality of water deteriorate under the changed climate scenario. The increase in such kind of losses also decreases the performance efficiency of turbines.

That is why the present study is an attempt to establish a relationship between some parameters affected by climate change and overall efficiency of turbines attached to hydropower plants.

The problem is introduced in Chap. 1 and methodology of the procedure adopted to quantify the climatic impact on turbine performance is described in Chap. 4. A brief description about hydropower plants, types of turbines and types of losses which reduces the turbine efficiency was discussed in Chap. 3. The cause and effect of climate change was delineated in Chap. 2. The results of the case studies and model validation were depicted in Chap. 5 followed by the conclusions that were derived from the present investigation were shown in the last chapter (Chap. 6).

References

Kumar D, Sarkar S (2016) A review on the technology, performance, design optimization, reliability, techno-economics and environmental impacts of hydrokinetic energy conversion systems. Renew Sustain Energy Rev 58:796–813

Majone B, Villa F, Deidda R, Bellin A (2016) Impact of climate change and water use policies on hydropower potential in the south-eastern Alpine region. Sci Total Environ 543:965–980

Wang X, Zhang J, Shahid S, Guan E, Wu Y, Gao J, He R (2016) Adaptation to climate change impacts on water demand. Mitig Adapt Strat Glob Change 21(1):81–99

Acknowledgements

The authors would like to take this opportunity to convey their gratitude for their colleagues, friends and family, without the cooperation from them this investigation and consequent preparation of the manuscript may never be completed.

The authors would also like to thank the reviewers, editors and publishers for taking the pain to publish this book.

Throughout the book many references were cited. The authors have collected data and information from various sources. Authors would like to acknowledge the publishers of all the references and sources that were used during the investigation phase and corresponding manuscript preparation phase.

Last but not least, the authors would be grateful if the mistakes that were inadvertently incorporated in the manuscript can be identified and communicated to the authors, so that they can be rectified.

Agartala, India   Uttam Roy
2016   Mrinmoy Majumder

Contents

Chapter 1
Introduction

Abstract The quality and quantity of water gets effected due to the change in the climatic pattern. The performance efficiency of turbines attached to hydropower plants are directly effected by the quality and quantity of water in which the runner is either submerged or rotates in of the water. That is why climate change also impacts the performance of hydro-turbines. The present study is an attempt to quantify the relationship between the climatic impact and turbine performance.

Keywords Hydropower · Hydro turbines · Climate change

Hydro-turbines are turbines used in hydropower plants.

The rising demand for electricity in the rapidly urbanizing world has induced scarcity in energy supply. That is why the search for suitable alternative to conventional energy sources is popular field of research post 1980s.

The hydro-energy or "energy from water" is referred as the most reliable but inexpensive form of renewable energy which has the maximum potential to replace conventional fuel sources.

The efficiency of turbines largely controls the efficacy of the power conversion procedure that takes place in the power house of hydro-power plant (De Jaeger et al. 1994).

The performance efficiency (Gaiser et al. 2016; Kang 2016) of turbines depends largely on the shape and size of the turbine along with the efficiency of the materials with which it was developed.

The wear and tear caused by various types of frictions (Liu et al. 2016) and corrosion and erosion from the flowing water is found to be few of the major reasons for which the efficiency of turbines gets reduced. The caveats formed from water bubbles or impulse from the water jets may also reduce the lifetime of the turbines installed in the hydropower plants. The suspended solids can also inflict tears in the turbine blades (Mo et al. 2016).

The erosive power of the flowing water depends on the head (potential energy) and rate of flow (kinetic energy) (Mahdipoor et al. 2016). Again the corrosive power will depend on the quality of water. Both head and flow depends on climatic

U. Roy and M. Majumder, *Impact of Climate Change on Small Scale Hydro-turbine Selections*, SpringerBriefs in Energy,
DOI 10.1007/978-981-287-239-5_1

variables (Xu et al. 2016). The quality parameters indirectly depend on climatic parameters like temperature and precipitation (François et al. 2016).

The loss in the turbine performance will increase the maintenance cost of the equipment. The operational cost will also increase if cavitations loss gets increased. The overall operation cost of the hydropower plant will thus be incremented. That is why; the quantification of the impact of the climatic parameter on efficiency of the turbines is an important aspect which can mitigate both the loss of performance and economic liability of the power plant (Arce et al. 2002; Zhang et al. 2016).

1.1   Objective of the Study

The main objective of the present study is to estimate the vulnerability of hydro-turbines installed in hydropower plants to climatic vulnerabilities with the help of a media which will be both cognitive and objective in nature.

The study aims to identify the priority parameters and the priority values of the variables.

The study also tries to establish an indicator which can be used as a representative of the climatic vulnerabilities to hydro-turbines installed in hydro power plants.

1.2   Brief Methodology

The present investigation adopted the advantages of Multi Criteria Decision Making methods (Ziaei et al. 2016) to identify and estimate the priority and priority value of the parameters. The method selected was objective, logical and compares one option with other based on decision goals.

The indicator was developed with the help of the priority value and priority parameters.

The indicator was estimated with the help of a new variant of Artificial Neural Network; Group Method of Data Handling (GMDH) (Pourkiaei et al. 2016), which also makes the indicator cognitive and platform flexible.

The indicator was used to measure the vulnerability of hydro-turbine for different power plants situated all over the World to climate change. In total six power plants were evaluated with the help of the indicator based neural network model.

References

Arce A, Ohishi T, Soares S (2002) Optimal dispatch of generating units of the Itaipú hydroelectric plant. IEEE Trans Power Syst 17(1):154–158

De Jaeger E, Janssens N, Malfliet B, Van De Meulebroeke F (1994) Hydro turbine model for system dynamic studies. IEEE Trans Power Syst 9(4):1709–1715

François B, Hingray B, Raynaud D, Borga M, Creutin JD (2016) Increasing climate-related-energy penetration by integrating run-of-the river hydropower to wind/solar mix. Renew Energy 87:686–696

Gaiser K, Erickson P, Stroeve P, Delplanque JP (2016) An experimental investigation of design parameters for pico-hydro Turgo turbines using a response surface methodology. Renew Energy 85:406–418

Kang SH (2016) Design and preliminary tests of ORC (organic Rankine cycle) with two-stage radial turbine. Energy 96:142–154

Liu X, Luo Y, Wang Z (2016) A review on fatigue damage mechanism in hydro turbines. Renew Sustain Energy Rev 54:1–14

Mahdipoor MS, Kevorkov D, Jedrzejowski P, Medraj M (2016) Water droplet erosion mechanism of nearly fully-lamellar gamma TiAl alloy. Mater Des 89:1095–1106

Mo Z, Xiao J, Wang G (2016) Numerical research on flow characteristics around a hydraulic turbine runner at small opening of cylindrical valve. Math Probl Eng 2016

Pourkiaei SM, Ahmadi MH, Hasheminejad SM (2016) Modeling and experimental verification of a 25 W fabricated PEM fuel cell by parametric and GMDH-type neural network. Mech Ind 17 (1):105

Xu Q, Li W, Lin Y, Liu H, Yajing Gu (2016) Investigation of the performance of a stand-alone horizontal axis tidal current turbine based on in situ experiment. Ocean Eng 113:111–120

Zhang Y, Qian Z, Ji B, Wu Y (2016) A review of microscopic interactions between cavitation bubbles and particles in silt-laden flow. Renew Sustain Energy Rev 56:303–318

Ziaei M, Sajadi MS, Tavakoli MM (2016) The performance improvement of water pump manufacturing system via multi-criteria decision-making and simulation (a case study: Iran Godakht Company). Int J Prod Qual Manage 17(1):1–15

Chapter 2
Climate Change and Its Impacts

Abstract There are two types of causes which can induce change in the climate of a region, namely, natural and human induced causes. The change in climate due to volcanic activity, changes in solar output or change in the earth's orbit around the sun can are examples of natural causes. The human induced causes include but not limited to burning of fossil fuels and conversion of forest or agricultural lands. As a result of the change in regular pattern of climate both water and energy sector has been severely effected. The effect is dominant on the urban as well as rural environment.

Keywords Climate impact · Climate change · Natural resources

The climate change impact on natural resources and energy sources are well documented in many related literatures and reports (Poudyal et al. 2016).

The hydropower plants being water based energy resources has been found to be directly impacted by the change in climatic pattern of the precipitation and temperature (DeNooyer et al. 2016; Gaudard et al. 2016; Koch et al. 2016).

The change in climatic pattern can be attributed to various causes which can be classified under natural and anthropogenic influences.

In short it can be explained that due to the increase in Green House Gases (GHG) like Carbon di-oxide, Methane etc. the atmospheric temperature has also increased as all the gases under this class is an absorbent of heat. The contribution from industrial and agricultural industries along with auto-mobiles the density of GHG gases has increased beyond limit. As an effect the climatic pattern has also changed. Some of the impacts of climate change can be found in Fig. 2.1.

The impact of climate change on energy resources and more importantly water based energy resources has been explained and identified in Iliadis and Gnansounou (2016); Huang et al. (2016), Ludwig et al. (2016), Molden et al. (2016) etc.

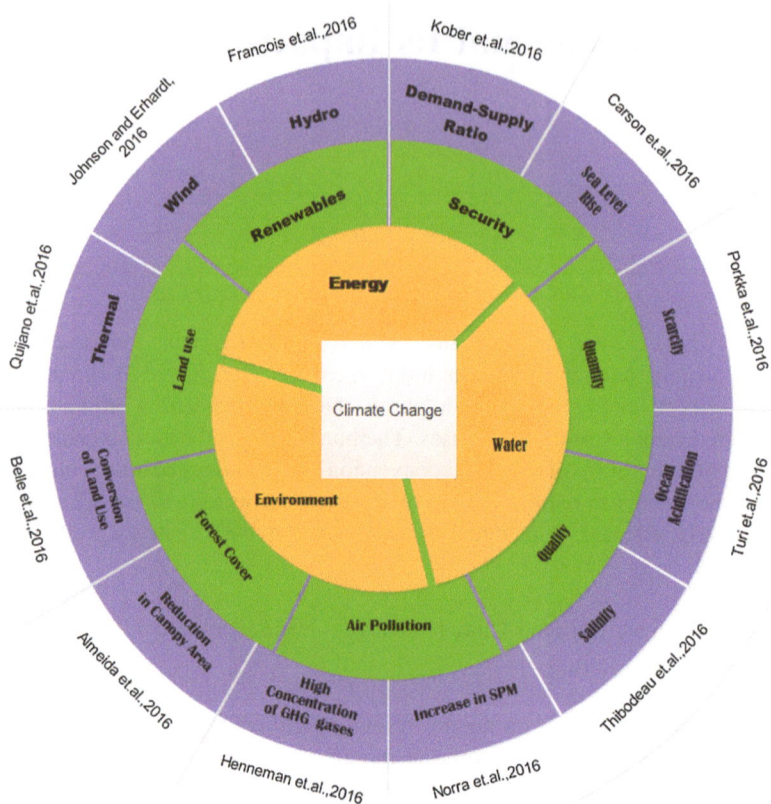

Fig. 2.1 Figure showing the impacts of climate change

The Inter-governmental Panel on Climate Change (IPCC) was established to work about the impact of climate change. The Panel has simulated the future scenario based on the change in climatic pattern, socio-economic stability and increase in urban population.

In total four scenarios were made viz, A1, A2, B1 and B2. In the first two scenarios, world is taken as globally united. But in case of A1 scenario situation more economical than environmental and in B1 scenario the situation is opposite. The only difference between A1 and A2 or B1 and B2 is that in the latter world is regionally divided. Figure 2.2 shows a schematic diagram of the four main scenario and the way it simulates the situation that may or may not exist in the near future.

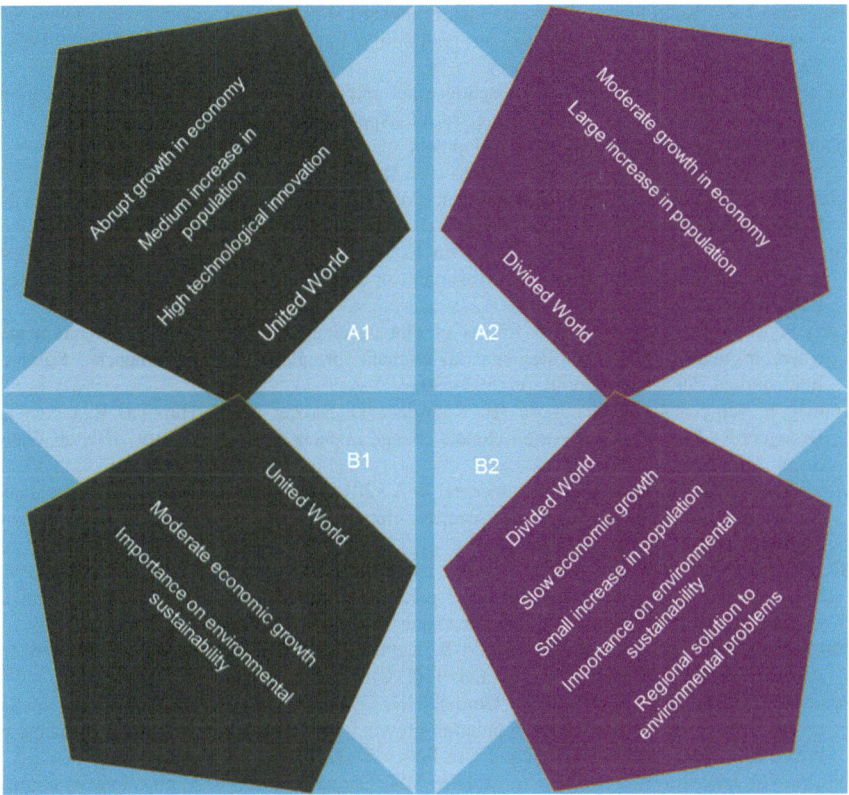

Fig. 2.2 Figure showing the four main IPCC scenarios of climate change

References

Almeida Castanho AD, Galbraith D, Zhang K, Coe MT, Costa MH, Moorcroft P (2016) Changing Amazon biomass and the role of atmospheric CO_2 concentration, climate, and land use. Global Biogeochem Cycles

Belle S, Verneaux V, Millet L, Etienne D, Lami A, Musazzi S, Reyss J-L, Magny M (2016) Climate and human land-use as a driver of Lake Narlay (Eastern France, Jura Mountains) evolution over the last 1200 years: implication for methane cycle. J Paleolimnol 55(1):83–96

Carson M, Köhl A, Stammer D, Slangen ABA, Katsman CA, van de Wal RSW, Church J, White N (2016) Coastal sea level changes, observed and projected during the 20th and 21st century. Climatic Change 134(1–2):269–281

DeNooyer TA, Peschel JM, Zhang Z, Stillwell AS (2016) Integrating water resources and power generation: the energy–water nexus in Illinois. Appl Energy 162:363–371

Francois B, Borga M, Creutin J-D, Hingray B, Raynaud D, Sauterleute J-F (2016) Complementarity between solar and hydro power: sensitivity study to climate characteristics in Northern-Italy. Renew Energy 86:543–553

Gaudard L, Gabbi J, Bauder A, Romerio F (2016) Long-term uncertainty of hydropower revenue due to climate change and electricity prices. Water Resour Manag 1–19

Henneman LRF, Rafaj P, Annegarn HJ, Klausbruckner C (2016) Assessing emissions levels and costs associated with climate and air pollution policies in South Africa. Energy Policy 89:160–170

Huang W, Ma D, Chen W (2016) Connecting water and energy: assessing the impacts of carbon and water constraints on China's power sector. Appl Energy

Iliadis NA, Gnansounou E (2016) Development of the methodology for the evaluation of a hydro-pumped storage power plant: swiss case study. Energy Strateg Rev 9:8–17

Johnson DL, Erhardt RJ (2016) Projected impacts of climate change on wind energy density in the United States. Renew Energy 85:66–73

Kober T, Falzon J, van der Zwaan B, Calvin K, Kanudia A, Kitous A, Labriet M (2016) A multi-model study of energy supply investments in latin America under climate control policy. Energy Econ

Koch F, Reiter A, Bach H (2016) Effects of climate change on hydropower generation and reservoir management. In: Regional assessment of global change impacts. Springer International Publishing, Berlin, pp. 593–599

Ludwig F, van Schaik H, Matthews JH, Rodriguez D, Bakker MHN, Huntjens P, Lexén K, Droogers P (2016) Perspectives on climate change impacts and water security. Handb Water Secur 139

Molden DJ, Shrestha AB, Nepal S, Immerzeel WW (2016) Downstream implications of climate change in the Himalayas. In: Water security, climate change and sustainable development. Springer, Singapore, pp. 65–82

Norra S, Yu Y, Dietze V, Schleicher N, Fricker M, Kaminski U, Chen Y, Stüben D, Cen K (2016) Seasonal dynamics of coarse atmospheric particulate matter between 2.5 μm and 80 μm in Beijing and the impact of 2008 Olympic Games. Atmos Env 124:109–118

Porkka M, Gerten D, Schaphoff S, Siebert S, Kummu M (2016) Causes and trends of water scarcity in food production. Env Res Lett 11(1):015001

Poudyal NC, Elkins D, Nibbelink N, Cordell HK, Gyawali B (2016) An exploratory spatial analysis of projected hotspots of population growth, natural land loss, and climate change in the conterminous United States. Land Use Policy 51:325–334

Quijano JC, Jackson PR, Santacruz S, Morales VM, García MH (2016) Implications of climate change on the heat budget of lentic systems used for power station cooling: case study Clinton Lake, Illinois. Env Sci Technol

Thibodeau B, Bauch D (2016) The impact of climatic and atmospheric teleconnections on the brine inventory over the Laptev Sea shelf between 2007 and 2011. Geochem Geophys Geosyst

Turi G, Lachkar Z, Gruber N, Münnich M (2016) Climatic modulation of recent trends in ocean acidification in the California Current System. Env Res Lett 11(1):014007

Chapter 3
Hydropower Plants

Abstract The turbines used in hydropower plant can be classified into high, medium and low head turbines. The specific speed and head requirement of turbines helps to classify them into the above classes. The turbine efficiency is reduced due to losses like corrosion, erosion, fatigue and friction. This chapter deals with various type of turbines and the losses which decreases their performance efficiency.

Keywords Hydropower plant · Turbines · Vulnerabilities

The hydropower plants are installed to produce energy from running water. There are various types and forms of hydropower plant available all over the world but their working principle same as depicted in Fig. 3.1.

The energy of flowing water is converted to rotate turbines which in turn rotate the rotor of the generator device and according to faraday's law an electro-motive force is created which produce current when the circuit is completed.

The energy conversion procedure follows the Bernoulli's Law which states that "If no energy is added to the system, an increase in velocity is accompanied by a decrease in density and/or pressure. The law is directly related to the principle of conservation of energy." The pressure head of water stored is converted into velocity head. In case of run off the river hydropower plant, the kinetic energy of the flowing water is directly utilized to produce energy by rotating turbine and the alternator.

The turbine attached to the alternator can be of different types based on head required and specific speed. Figure 3.2 shows different types of turbine used in hydropower plants.

3.1 Turbine Vulnerabilities

There are various types of turbines used in different types of hydropower plant.

Figure 3.3 shows the type and form of losses observed in turbine used in hydropower plants.

© The Author(s) 2016   9
U. Roy and M. Majumder, *Impact of Climate Change on Small Scale*
Hydro-turbine Selections, SpringerBriefs in Energy,
DOI 10.1007/978-981-287-239-5_3

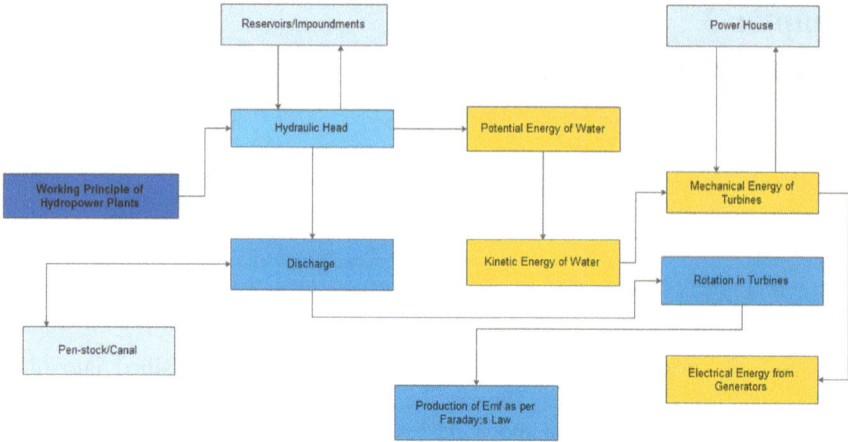

Fig. 3.1 Figure showing the working principle of hydropower plants

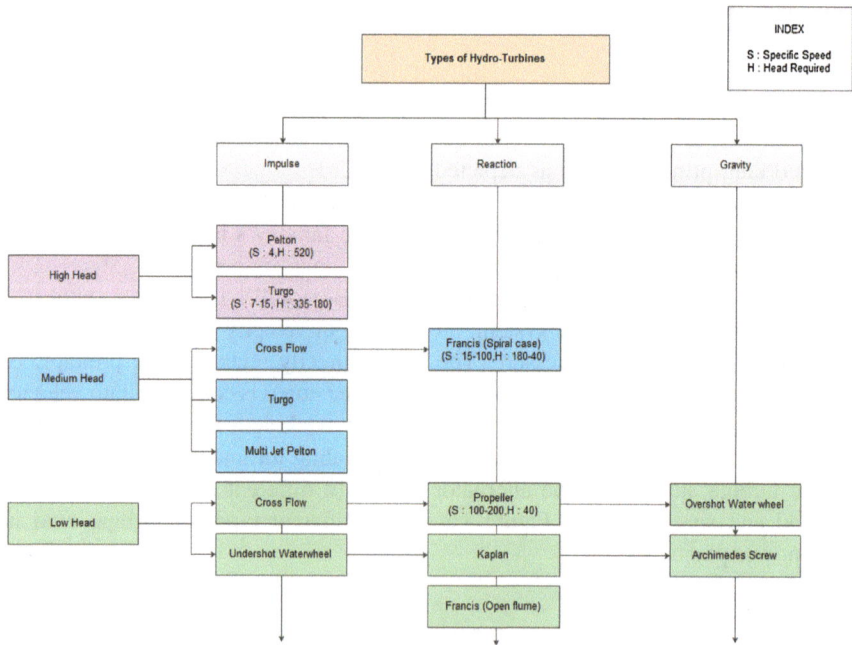

Fig. 3.2 Figure showing the different types of turbines utilized in hydropower plant

The factors like corrosion and erosion, fatigue along with blade friction reduce the efficiency of turbines attached to hydropower plant. The impact of the losses described in Fig. 3.3 is also expressed in various literatures like Benzon et al. (2016); Gohil and Saini (2016); Huang (2016) etc.

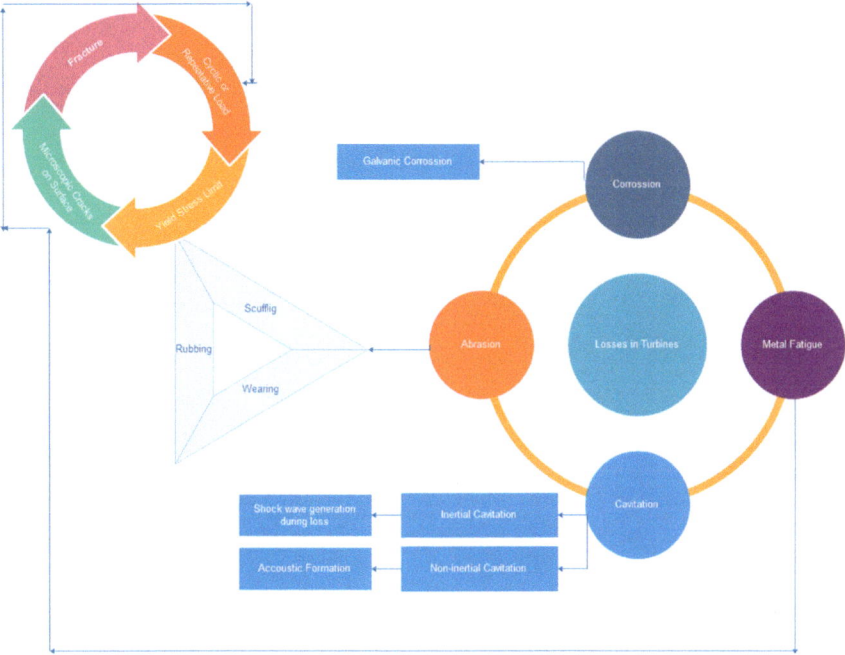

Fig. 3.3 Figure showing different types of losses incurred on a turbine

References

Benzon DS, Aggidis GA, Anagnostopoulos JS (2016) Development of the turgo impulse turbine: past and present. Appl Energy 166:1–18

Gohil PP, Saini RP (2016) Numerical study of cavitation in Francis turbine of a small hydro power plant. J Appl Fluid Mech 9(1):357–365

Huang X (2016) Developing corrosion prevention coating solutions for the canadian SCWR concept. JOM 68(2):480–484

Chapter 4
Methodology

Abstract The present investigation tries to estimate the relationship between climatic impact and turbine efficiency. In this regard seven parameters were identified as input and the ratio of weighted summation of the beneficiary and non-beneficiary parameters are taken as output. This ratio is directly proportional to turbine performance. The weights of the parameters were identified with the help of three MCDM methods; AHP, ANP and MACBETH. The relationship between input and output parameters is established by the GMDH technique. The selected model from the twenty four models developed in this aspect was applied to quantify the climatic impact on the turbines installed in six different places located in different continents. The impact of climate change is depicted by the output from another model developed to map climatic parameters to the model input parameters.

Keywords Multi criteria decision making · Group method of data handling · Turbine vulnerability

The present investigation utilized two methods in a cascaded formation.

The MCDM method Analytical Hierarchy Process (AHP), Analytical Network Process (ANP) along with MACBETH and a new ANN based method Group Method of Data Handling (GMDH).

Tables 4.1 and 4.2. has depicted the strength and weakness of each of this techniques.

The entire methodology adopted to estimate turbine vulnerabilities towards climate change is given in Fig. 4.1.

The decision making methodology was discussed in the next section.

4.1 Application of MCDM Methods

The application of MCDM comprises of Selection of Criteria and alternatives, rating the alternatives with respect to the criteria and lastly the aggregation of the rating will give the priority value of the input variables.

U. Roy and M. Majumder, *Impact of Climate Change on Small Scale Hydro-turbine Selections*, SpringerBriefs in Energy, DOI 10.1007/978-981-287-239-5_4

Table 4.1 Table showing the strength, weakness and purpose of using the MCDM methods

	AHP	ANP	MACBETH
Strength	Flexible	Network or bi-directional structure of decision making	Flexible
	Easy to use	Compares alternatives with respect to criteria and vice versa	Ease of use
	Relative way of comparison	Flexible	The fuzziness of rating for similarly important variables were reduced by the introduction of deductive way of rating the comparison instead of rating based on ratio
	Rating based on crisp nine point scale	Ease of use	Considers both qualitative and quantitative variables
	Considers both qualitative and quantitative variables	Considers both qualitative and quantitative variables	
Weakness	Hierarchical or uni directional structure of decision making	Fuzziness or improper rating in cases of rating similarly important variables	The resultant rating often becomes negative
	Fuzziness or improper rating in cases of rating similarly important variables		
Why used	The priority value of the parameters were estimated by this method	The priority value of the parameters were estimated by this method	The priority value of the parameters were estimated by this method
Application	Qdais et al. (2016)	Razavi Toosi et al. (2016)	Marques and Silve (2016)

In total 7 variables were selected as alternative. There were three criteria viz hydraulic head, flow and quality of water. The Climatic and economic impact along with performance efficiency of the turbine was taken as super criteria.

The hydraulic head, flow and quality of water was compared with respect to climatic and economic impact and performance efficiency of the turbines.

The alternatives: Specific Speed of the turbine, Corrodibility of turbine blades, Material Life Time of the runner, Shaft Efficiency and Installation, Operational and Maintenance Cost; were selected to represent the performance efficiency of hydro-turbines.

The ranking of the alternatives as per their importance were performed with the help of information retrieved from literature survey about the decision objective.

Table 4.2 Table showing the strength, weakness and justification for using GMDH method in the present investigation

	GMDH
Strength	The method considers multiple training algorithms to solve the same problem and select the output from the algorithm which have resulted in optimal performance of the model
	Considers nonlinearity
	Flexible
	Ease of use
	Considers both multiplier input single output and multiple input multiple output
Weakness	As complexity of the model depends on the number of inputs so more the number of input more will be the complexity of the model
	The architecture of the model is selected based on the problem complexity and there is no scope of voluntary modifications
Application	Zhang (2016)

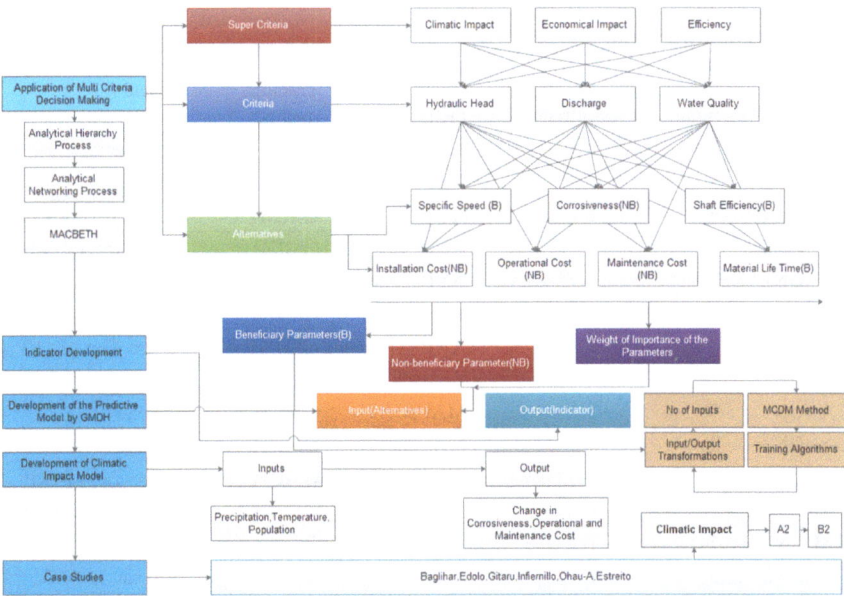

Fig. 4.1 Figure showing the schematic diagram of methodology followed to achieve the present objective

The method AHP, ANP and MACBETH was used to aggregate the rating received by each of the alternative with respect to the criteria considered. The result from the aggregation yielded the priority value of the selected variables.

In case of ANP the criteria were compared with each other with respect to the alternative also and the ratings for the alternative for the MACBETH method was performed by the subtraction rather than ratio method.

4.2   Application of GMDH

In the next phase of the study neural network models were developed where all the selected alternatives were taken as input and the ratio of weighted summation of the beneficiary and non-beneficiary parameters were taken as output. The parameters were considered beneficiary if the increase in magnitude of the variable increases the turbine efficiency and vice versa.

In total twenty four models were developed. In the models the weights used in the output parameter was derived from AHP, ANP and MACBETH method. The transformation of the output was also executed with the help of Arc Tangent function and numbers of inputs were varied between seven and three. The training algorithm was also changed between GMDH and Quick Combinatorial method. Table 4.3 depicts the characteristic of the twenty four models developed for the present study objective.

Table 4.3   Table showing the characteristics of the models developed for the present investigation

Model no.	No. of input	No. of output	Priority method	Training method	Data transformation
7AHGN1	**7**	**1**	**AHP**	**GMDH**	**None**
3AHGN2	3	1	AHP	GMDH	None
7AHGAO1	**7**	**1**	**AHP**	**GMDH**	**Arc Tan of O/P**
3AHGAO2	3	1	AHP	GMDH	Arc Tan of O/P
7ANGN1	7	1	ANP	GMDH	None
3ANGN2	3	1	ANP	GMDH	None
7ANGAO1	**7**	**1**	**ANP**	**GMDH**	**Arc Tan of O/P**
3ANGAO2	3	1	ANP	GMDH	Arc Tan of O/P
7ANGN1	7	1	MACBETH	GMDH	None
3ANGN2	3	1	MACBETH	GMDH	None
7ANGAO1	7	1	MACBETH	GMDH	Arc Tan of O/P
3ANGAO2	3	1	MACBETH	GMDH	Arc Tan of O/P
7AHCN1	7	1	AHP	C	None
3AHCN2	3	1	AHP	C	None
7AHCAO1	7	1	AHP	C	Arc Tan of O/P
3AHCAO2	3	1	AHP	C	Arc Tan of O/P
7ANCN1	7	1	ANP	C	None
3ANCN2	3	1	ANP	C	None
7ANCAO1	7	1	ANP	C	Arc Tan of O/P
3ANCAO2	3	1	ANP	C	Arc Tan of O/P
7ANGN1	7	1	MACBETH	C	None
3ANGN2	3	1	MACBETH	C	None
7ANGAO1	7	1	MACBETH	C	Arc Tan of O/P
3ANGAO2	3	1	MACBETH	C	Arc Tan of O/P

The best performing model was selected with the help of Mean Absolute Error and Correlation Coefficient and the resultant performance index (PI) which is directly proportional to Correlation Coefficient and inversely proportional to mean absolute error. The weightage for testing phase PI was more than the training phase PI.

After the best model was selected, second level of validation was conducted by comparing the model output for the best three models among the forty eight models considered with the help of Covariance, Mean Relative Error and Standard Deviation in between the predicted and desired output.

Here also an Equivalent Performance Index (EPI) was developed so that performance can be represented in a single point variable. The EPI was inversely proportional to all the metrics used and 60 % weightage was given to testing and the rest was kept for training phase EPI as importance of testing phase is more than training phase performance in case of neural network models (Valipour 2016; Ahmed et al. 2016)

The sensitivity analysis of the selected model was also performed.

4.3 Case Studies

The developed model was applied to find the performance of the turbines installed in the hydropower plants of Baglihar (Roberts 2016), Edolo (Sforna and Bertanza 2002), Gitaru (Muriithi 2006), Infiernillo (Labadie 2004), Ohau-A (Herath et al. 2011) and Estreito (de Souza 2008) located in India, Italy, Kenya, Mexico, New Zealand and Brazil respectively. Table 4.4 depicts the rank of each of the power plants compared to each other with respect to the input variables. The turbine types installed in the power plants were respectively Vertical Francis, Francis Pump-turbines and Kaplan turbine. The plants were installed respectively on 2008, 1984, 1978, 1965, 1979 and 2011.

The total capacity of the plants was 450, 1000, 225, 1000, 264 and 1087 respectively.

Table 4.4 Table showing the rank of each of the case study areas with respect to each other for the considered input variables

	Baglihar	Edolo	Gitaru	Infiernillo	Ohau-A	Estreito
Speed	2	1	5	6	4	3
Corrodibility	6	4	1	5	3	2
Material life time	2	4	3	6	5	1
Shaft efficiency	1	2	4	6	5	3
Install cost (IC)	2	3	5	6	4	1
Operational cost	5	3	4	1	2	6
Maintenance cost	4	3	6	2	5	1

4.4   Impact Analysis of Climate Change

The developed model was also utilized to predict the performance efficiency of the turbines under changed climate scenario. The CMIP 5 model output was used to predict the future climate scenario. The magnitude of the input variables under changed climate scenario was estimated by new model developed for the present study. The model was developed with the help of GMDH method. The input to the model was the "change" in the climatic variables like precipitation and temperature along with the "change" in population and the output was the change in the magnitude of the input variables of the model developed for performance efficiency estimation. The magnitude of the top three important parameters among the seven considered variables were estimated with the help of the former model for different climatic scenario as retrieved from the CMIP 5 model for the selected case study area.

References

Ahmad Z, Crowley D, Marina N, Jha SK (2016) Estimation of biosurfactant yield produced by Klebseilla sp. FKOD36 bacteria using artificial neural network approach. Measurement 81:163–173

de Souza ACC (2008) Assessment and statistics of Brazilian hydroelectric power plants: dam areas versus installed and firm power. Renew Sustain Energy Rev 12(7):1843–1863

Herath I, Deurer M, Horne D, Singh R, Clothier B (2011) The water footprint of hydroelectricity: a methodological comparison from a case study in New Zealand. J Clean Prod 19 (14):1582–1589

Labadie JW (2004) Optimal operation of multireservoir systems: state-of-the-art review. J Water Resour Plann Manag 130(2):93–111

Marques M, Neves-Silva R (2016) Decision support system for energy savings and emissions trading in industrial scenarios. In: Intelligent decision technology support in practice. Springer International Publishing, Berlin, pp. 31–47

Muriithi EJ (2006) Developing small hydropower infrastructure in Kenya. In: 2nd small hydropower for today conference, "policy stimulating hydropower development". international network on small hydropower (IN-SHP)

Qdais HA, Hussam A (2016) Selection of management option for solid waste from olive oil industry using the analytical hierarchy process. J Mat Cycles Waste Manag 18(1):177–185

Razavi Toosi SL, Samani JMV (2016) Evaluating water management strategies in watersheds by new hybrid fuzzy analytical network process (FANP) methods. J Hydrol

Roberts A (2016) Will rivers become a cause of conflict, rather than co-operation, in South Asia? In: Water security, climate change and sustainable development. Springer, Singapore, pp. 155–162

Sforna M, Bertanza VC (2002) Restoration testing and training in Italian ISO. Power Syst IEEE Trans 17(4):1258–1264

Valipour M (2016) Optimization of neural networks for precipitation analysis in a humid region to detect drought and wet year alarms. Meteorol Appl 23(1):91–100

Zhang Y-J (2016) Research on carbon emission trading mechanisms: current status and future possibilities. Int J Global Energy Issues 39(1–2):89–107

Chapter 5
Result and Discussion

Abstract The results show that specific speed, O&M costs and corrosiveness of the turbine are the most important parameters which can influence the performance efficiency of the runners. The model prepared with all the seven inputs and AHP method along with transformed data of output becomes the optimally performing model among the twenty four models prepared for the study. The climate change impact on the selected power plants depicted that for most of the plant the worst and best performance was observed during the first thirty years of A2 and last thirty years of B2 respectively.

Keywords Analytical hierarchy process · CMIP 5 · Impact analysis

The results derived from MCDM technique is depicted in Tables 5.1, 5.2 and 5.3.

The performance metrics of the neural network models developed to predict the indicator is depicted in Table 5.4 and the metrics for the three most accurate models among the twenty four considered was depicted in Table 5.5.

The model performance for generation of data for the input parameters with respect to the values of the climatic parameters and population change was depicted in Table 5.6.

The vulnerability analysis of the case study for normal as well as climatic scenario is given in Tables 5.7 and 5.8 respectively.

The most important parameter was found to be Specific Speed of turbine among the beneficiary and operational and maintenance cost along with corrodibility of turbine blades among the non-beneficiary parameters with respect to the study objective and as per most of the MCDM methods applied in this regard and depicted in Tables 5.1, 5.2 and 5.3. The importance of Specific Speed, O&M and installation cost was well documented in contemporary literatures like Martins et al. (2016), Gao et al. (2016), Das et al. (2016) and Yang et al. (2016) respectively. The impact of corrosion on turbine performance is also cited in various literatures like Phillips and Staniewski (2016) and Ciubotariu et al. (2016).

The most optimal model was found to be 7AHGAO1 i.e., the model which was developed with the help of seven inputs, priority values determined by AHP

Table 5.1 Table showing the rank and priority value of the selected parameters as per AHP method

	Head	Flow	Quality	Priority value	Rank
Criteria weights	0.298701	0.513577	0.187721		
Speed	0.302267	0.043478	0.089744	0.159537	2
Corrodibility	0.302267	0.043478	0.089744	0.159537	2
Material life time	0.037783	0.076087	0.358974	0.145102	4
Shaft efficiency	0.060453	0.050725	0.119658	0.082035	7
Install cost (IC)	0.060453	0.076087	0.059829	0.084246	6
Operational cost	0.060453	0.101449	0.179487	0.127978	5
Maintenance cost	0.100756	0.304348	0.051282	0.241565	1

Table 5.2 Table showing the rank and priority value of the selected parameter as per ANP method

	Head	Flow	Quality	Priority value	Rank
Criteria weight	0.294766	0.501875	0.203359		
Speed	0.385675	0.369069	0.094595	0.318147	1
Corrodibility	0.055096	0.073814	0.378378	0.130233	3
Material life time	0.064279	0.073814	0.189189	0.094466	6
Shaft efficiency	0.096419	0.123023	0.054054	0.101155	5
Install cost (IC)	0.077135	0.052724	0.063063	0.062022	7
Operational cost	0.192837	0.184534	0.094595	0.168692	2
Maintenance cost	0.128558	0.123023	0.126126	0.125286	4

Table 5.3 Table showing the rank and priority value of the selected parameter as per MACBETH method

	Head	Flow	Quality	Priority value	Rank
Criteria weight	0.333333	0	−0.33333		
Speed	0.385675	0.369069	0.094595	0.160714	1
Corrodibility	0.055096	0.073814	0.378378	0.125	7
Material life time	0.064279	0.073814	0.189189	0.130952	6
Shaft efficiency	0.096419	0.123023	0.054054	0.14881	3
Install cost (IC)	0.077135	0.052724	0.063063	0.142857	4
Operational cost	0.192837	0.184534	0.094595	0.154762	2
Maintenance cost	0.128558	0.123023	0.126126	0.136905	5

Table 5.4 Table showing the performance metrics of the developed forty eight models for the present investigation

Model No.	No. of input	No. of output	Priority method	Training method	Data transformation	MAE (%) training	MAE (%) testing	Correlation (%) training	Correlation (%) testing	EPI	Rank
7AHGN1	**7**	**1**	**AHP**	**GMDH**	**None**	**1.92**	**1.94**	**99.93**	**99.71**	51.655	3
3AHGN2	3	1	AHP	GMDH	None	30.04	30.18	77.96	72.56	2.480	21
7AHGAO1	**7**	**1**	**AHP**	**GMDH**	**Arc Tan of O/P**	**0.75**	**1.03**	**99.97**	**99.92**	108.867	1
3AHGAO2	3	1	AHP	GMDH	Arc Tan of O/P	22.86	22.37	79.47	74.79	3.397	16
7ANGN1	7	1	ANP	GMDH	None	2.39	2.40	99.92	99.79	41.670	4
3ANGN2	3	1	ANP	GMDH	None	26.24	22.25	81.97	79.63	3.379	17
7ANGAO1	**7**	**1**	**ANP**	**GMDH**	**Arc Tan of O/P**	**1.10**	**1.33**	**99.91**	**99.82**	80.659	2
3ANGAO2	3	1	ANP	GMDH	Arc Tan of O/P	14.32	10.48	87.47	89.24	7.368	10
7ANGN1	7	1	MACBETH	GMDH	None	2.67	3.99	99.90	99.44	28.776	7
3ANGN2	3	1	MACBETH	GMDH	None	32.19	32.98	74.44	44.16	1.723	23
7ANGAO1	7	1	MACBETH	GMDH	Arc Tan of O/P	2.11	2.83	99.76	99.67	39.223	5
3ANGAO2	3	1	MACBETH	GMDH	Arc Tan of O/P	22.19	23.28	69.91	68.89	3.034	20
7AHCN1	7	1	AHP	C	None	19.79	16.86	90.24	89.77	4.989	13
3AHCN2	3	1	AHP	C	None	30.93	31.29	76.63	70.76	2.347	22
7AHCAO1	7	1	AHP	C	Arc Tan of O/P	7.93	6.44	97.42	96.84	13.796	8
3AHCAO2	3	1	AHP	C	Arc Tan of O/P	23.14	21.80	79.26	75.71	3.453	15

(continued)

Table 5.4 (continued)

Model No.	No. of input	No. of output	Priority method	Training method	Data transformation	MAE (%) training	MAE (%) testing	Correlation (%) training	Correlation (%) testing	EPI	Rank
7ANCN1	7	1	ANP	C	None	8.58	7.61	98.39	97.33	12.222	9
3ANCN2	3	1	ANP	C	None	25.25	21.86	82.36	79.88	3.483	14
7ANCAO1	7	1	ANP	C	Arc Tan of O/P	16.51	12.89	83.46	82.40	5.777	12
3ANCAO2	3	1	ANP	C	Arc Tan of O/P	14.48	10.71	87.32	88.63	7.211	11
7ANGN1	7	1	MACBETH	C	None	23.28	27.99	87.50	75.28	3.071	18
3ANGN2	3	1	MACBETH	C	None	34.37	34.49	61.24	54.18	1.655	24
7ANGAO1	7	1	MACBETH	C	Arc Tan of O/P	2.88	3.36	99.53	99.59	31.429	6
3ANGAO2	3	1	MACBETH	C	Arc Tan of O/P	22.44	22.95	68.90	70.47	3.071	19

Bold text indicate the top performing model

Table 5.5 Table showing the performance metrics of the best three models among the developed forty eight models

Model No.	7AHGAO1		7ANGAO1		7AHGN1	
	Training	Testing	Training	Testing	Training	Testing
MRE	−0.00083	0.004219	−0.00031	−0.00098	0.003076	0.001476
RMSE	1.94	3.29	2.89	1.94	4.77	4.24
Covar	0.046267	0.051012	0.060068	0.059557	0.150334	0.262088
STDEV	1.94	3.22	2.89	1.93	4.76	4.24
PI	0.200748713		0.197033369		0.107440562	
Rank	1		2		3	

Table 5.6 Table showing the performance metrics of the model developed to estimate input parameters of the model 7AHGAO1 with respect to climatic parameters and population change

Metrics	Training phase	Testing phase
Root mean square error (%)	2.86×10^{-12}	1.96×10^{-12}
Mean absolute error (%)	1.39×10^{-12}	1.21×10^{-12}
Correlation (%)	99.99	99.99

Table 5.7 Table showing the estimation of performance efficiency of the turbines installed in the selected case study area

Case study area	Indicator value	Rank
Baglihar	0.87461	1
Edolo	0.666608	2
Gitaru	0.552139	4
Infiernillo	0.186882	6
Ohau-A	0.433543	5
Estreito	0.570071	3

Table 5.8 Table showing the impact of climate change on the performance of turbines installed in the selected case study area

Baglihar		
A2	0.785257	5
A2	0.820449	3
A2	0.851133	1
B2	0.784551	6
B2	0.817972	4
B2	0.847779	2
Edolo		
A2	0.660553	6
A2	0.681939	5
A2	0.703334	3
B2	0.695302	4

(continued)

Table 5.8 (continued)

B2	0.719348	2
B2	0.743702	1
Gitaru		
A2	0.655045	5
A2	0.696343	2
A2	0.733088	1
B2	0.626203	6
B2	0.660874	4
B2	0.690664	3
Infiernillo		
A2	0.591502	6
A2	0.634388	4
A2	0.672664	2
B2	0.614405	5
B2	0.656312	3
B2	0.693709	1
Ohau-A		
A2	0.694671	4
A2	0.693679	5
A2	0.692688	6
B2	0.723878	3
B2	0.729895	2
B2	0.737374	1
Estreito		
A2	0.613124	6
A2	0.646936	4
A2	0.6781	2
B2	0.61595	5
B2	0.659614	3
B2	0.698926	1

method, output data was transformed by arc tangent function and GMDH was used as a training algorithm. The sensitivity analysis of the model was shown in Fig. 5.1. The models 7ANGAO1 and 7AHGN1 was the second and third most efficient model as per the performance index created for the present study (Figs. 5.2, 5.3 and 5.4).

The index also identified that among the six different hydropower plants located in various places of the World the Infiernillo HPP in Mexico was found to have the

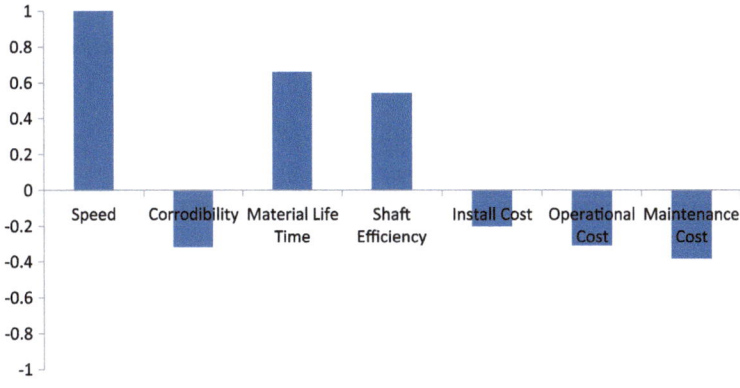

Fig. 5.1 Figure showing the sensitivity of selected input parameters for the present investigation

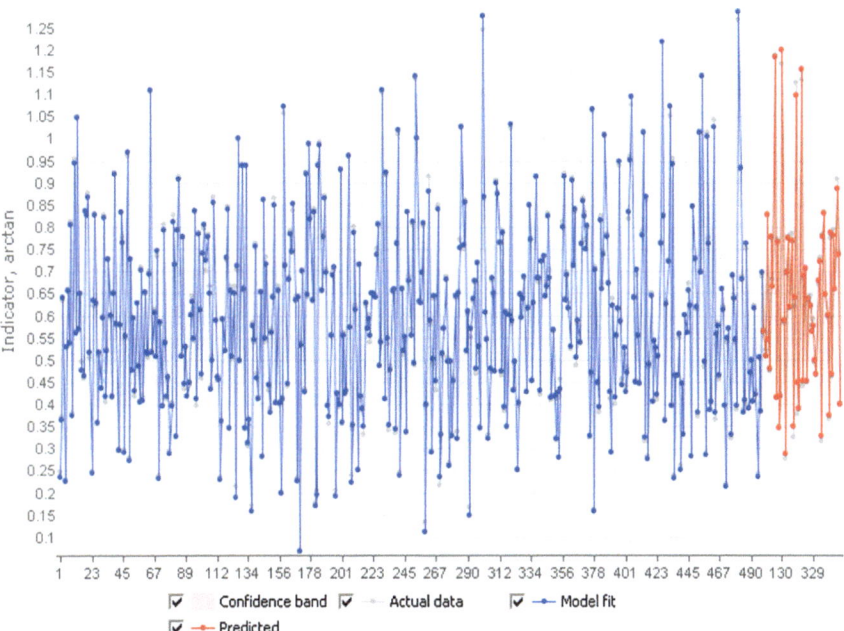

Fig. 5.2 Figure showing comparison between observed and predicted output for model No. 7AHGAO1

least and Baglihar in India have the highest performance efficiency among all the considered HPPs in the study. Marengo et al. (2013) and Zawahri (2009) has depicted similar importance of the respective power plants although in a different perspective.

Most of the turbines installed in hydropower plants were most vulnerable in the first thirty years of climate change as per IPCC A2 scenario and least vulnerable in the last thirty years of the B2 scenario. The turbines installed in Ohau-A power plant was found to be most and least vulnerable in the last thirty years of the A2 and B2 climate change scenarios respectively. Rest of the turbines installed in the power plants was found to have most severe climatic vulnerability on last thirty years of A2 and first thirty years of B2.

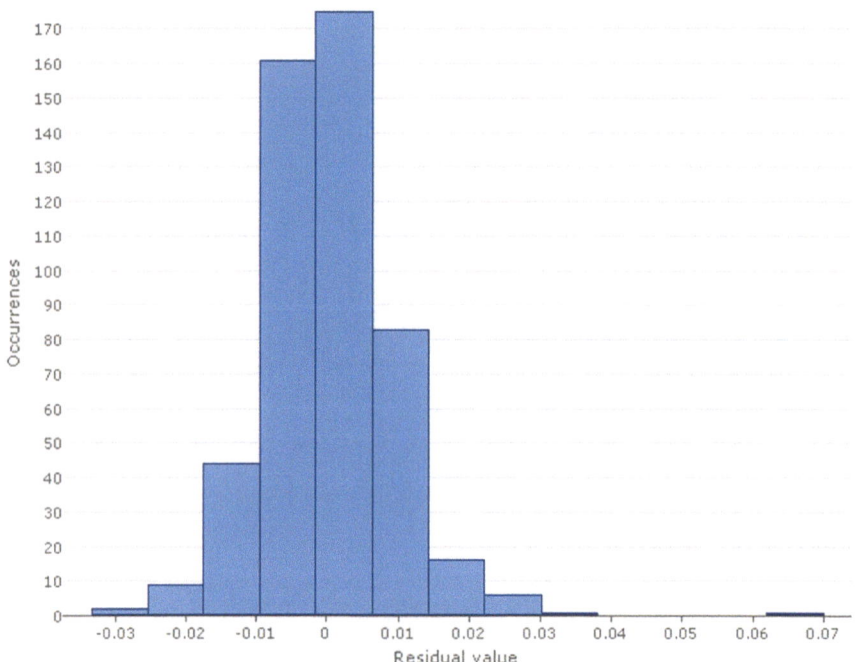

Fig. 5.3 Figure showing distribution of residuals from observed and predicted output for model No. 7AHGAO1

Model Equation # 5.1. for Model No. 7AHGAO1

$Y1 = 0.000699537 + N9*N2*0.534449 - N9^2*0.532755 + N2*0.997677$

$N2 = -0.00229556 - N192*N3*1.38623 + N192^2*0.703881 + N3*1.00458 + N3^2*0.679298$

$N3 = -0.00623896 + N424*0.0752731 + N424*N4*0.223914 - N424^2*0.172138 + N4*0.949995 - N4^2*0.0715399$

$N4 = 0.00117451 - N334*0.0530135 - N334*N5*0.614104 + N334^2*0.343934 + N5*1.04813 + N5^2*0.272139$

$N5 = -0.00139184 + N348*0.0712016 + N348*N6*0.379221 - N348^2*0.24478 + N6*0.934849 - N6^2*0.138612$

$N6 = 0.0911577 - N536*0.381579 - N536*N7*0.164354 + N536^2*0.39098 + N7*1.08148 + N7^2*0.0150003$

$N7 = 0.000205927 - N120*0.107688 - N120^2*0.000881168 + N8*1.10794$

$N8 = 0.00500828 - N316*0.0981996 + N316*N9*0.189443 + N9*1.07765 - N9^2*0.169692$

$N316 = -0.535156 + N474*1.32567 - N474^2*0.408176 + N479*0.835497$

$N120 = -0.0115022 + N167*0.798667 + N167*N202*0.330993 - N167^2*0.345643 + N202*0.231213$

$N202 = 0.188991 - N479*0.699937 + N479*N274*1.13051 + N479^2*0.0710206 + N274*1.071 - N274^2*0.668397$

$N274 = -0.533001 + N432*1.22999 + N432*N511*0.401512 - N432^2*0.387584 + N511*0.676051$

$N511 = 0.187404 + N530*0.734717 + N530*N533*0.51415 - N533*1.53635 + N533^2*1.95605$

$N432 = 0.261187 - N510*0.787328 + N510*N523*1.57809 + N510^2*0.654486$

$N167 = -0.00688555 + N204*0.997775 - N204*N387*0.294121 + N387^2*0.313863$

$N387 = -8.05257e-06 + N462*0.667824 + N462*N478*0.873362 - N462^2*0.311471 - N478*0.224025 + N478^2*0.357099$

$N462 = -1.11812 + N514*1.86928 - N514*N523*1.44081 + N523*1.86691$

$N204 = -0.193597 + N492*0.814421 + N492*N493*0.707408 - N492^2*0.252069 + N493^2*0.394451$

$N493 = -0.10859 - \text{"Maintenance Cost"}*N531*1.06731 + \text{"Maintenance Cost"}^2*0.216838 + N531*1.60172$

$N536 = -6.73117e-05 + N540^2*0.839599 + N541^2*0.829372$

$N541 = 0.539384 + \text{"Shaft Efficiency"}*0.215897 - \text{"Shaft Efficiency"}*\text{"Install Cost(IC)"}*0.205049$

$N540 = 0.711803 - \text{"Operational Cost"}*0.378514 + \text{"Operational Cost"}^2*0.211855$

$N348 = -0.072965 + N412*0.549764 + N412*N527*1.31053 - N412^2*0.322447$

N412 = -0.25636 + N505*1.29164 + N505*N510*0.527781 - N505^2*0.53943 - N510*0.25823 + N510^2*0.677428

N334 = 0.0512974 - N504*0.543962 + N504*N474*0.9458 + N504^2*0.732182 + N474*0.496781 - N474^2*0.0975204

N474 = -1.01496 + N519*1.56118 - N519*N523*1.2847 + N519^2*0.194785 + N523*1.78812

N424 = -0.335074 + N499*0.960057 + N499^2*0.0545601 + N533^2*0.941262

N192 = -0.0218344 + N267*1.0443 - N267*N337*0.044853 - N267^2*0.270089 + N337^2*0.304629

N337 = 0.0250381 + N453*0.587553 + N453*N478*0.861997 - N453^2*0.220534 - N478*0.27681 + N478^2*0.420827

N453 = -0.963782 + N514*1.62176 - N514*N521*1.04011 + N521*1.61686

N514 = 0.477845 - Corrodibility*0.22126 - Corrodibility*"Material Life Time"*0.10149 + Corrodibility^2*0.0473828 + "Material Life Time"*0.588657 - "Material Life Time"^2*0.203383

N267 = -0.367767 + N479*N498*0.679814 + N479^2*0.388315 + N498*1.44773 - N498^2*0.75964

N498 = -0.352291 + Speed*N524*0.363708 + Speed^2*0.158133 + N524*1.86462 - N524^2*0.907495

N479 = 0.615344 - N515*1.68502 + N515*N531*2.66365 + N515^2*0.945338 - N531*0.516683

N9 = -0.000150466 - N241*0.173527 - N241*N10*2.7295 + N241^2*1.45391 + N10*1.172 + N10^2*1.27395

N10 = -0.0101502 + N384*0.103525 - N384*N11*1.03508 + N384^2*0.446427 + N11*0.930126 + N11^2*0.558259

N11 = 0.00541457 + N14*0.986481 + N14*N16*97.347 - N14^2*48.7752 - N16^2*48.5598

N16 = -0.00342048 + N438*0.0978984 + N438*N18*0.473727 - N438^2*0.314747 + N18*0.924354 - N18^2*0.176737

N438 = 0.356043 - N506*0.90326 + N506*N531*2.37408 + N506^2*0.449848 - N531*0.400716 + N531^2*0.0189271

N531 = 0.821421 - Corrodibility*0.22452 - "Operational Cost"*0.383731 + "Operational Cost"^2*0.21781

N14 = 0.00247069 + N351*N18*0.870811 - N351^2*0.379876 + N18*0.991569 - N18^2*0.479072

N18 = -0.0323596 + "Operational Cost"*0.117178 - "Operational Cost"*N28*0.0514762 - "Operational Cost"^2*0.0918814 + N28*1.04187 - N28^2*0.01428

N28 = -0.00325282 - N407*0.0182561 - N407*N42*1.11985 + N407^2*0.594342 + N42*1.02139 + N42^2*0.519704

N42 = -6.87232e-05 + N260*0.236156 + N260*N74*1.24268 - N260^2*0.612556 + N74*0.757482 - N74^2*0.617227

N74 = 0.126287 - N485*0.461341 + N485*N144*0.471824 + N485^2*0.17854 + N144*1.04004 - N144^2*0.301687

N144 = -0.0207509 + N211*0.645204 - N211*N276*0.018375 + N276*0.401932

N276 = -0.415473 + N441*0.716644 + N441*N518*0.414392 + N441^2*0.0172637 + N518*0.723476

N518 = -0.677102 + N530*1.22568 - N530^2*0.170651 + N538*0.972584 + N538^2*0.0736534

N530 = 0.39312 + "Material Life Time"*0.609493 - "Material Life Time"*"Install Cost(IC)"*0.244501 - "Material Life Time"^2*0.148037

N441 = 0.189819 + Speed*0.350219 + Speed*N504*0.317943 - Speed^2*0.17186 - N504*0.190149 + N504^2*0.797471

N504 = 0.962929 - Corrodibility*0.249708 - "Maintenance Cost"*0.644945 + "Maintenance Cost"^2*0.234581

N211 = -0.118749 - N478*0.15594 + N478*N492*0.668003 + N478^2*0.510171 + N492*0.774256 - N492^2*0.224082

N478 = -0.0749835 + N510*N533*1.7412 + N533*0.0917495

N485 = 0.131236 - N510*0.424738 + N510*N534*2.37702 - N534^2*0.348116

N260 = 0.194991 - N460*0.907361 + N460*N282*0.875804 + N460^2*0.282017 + N282*1.24031 - N282^2*0.597057

N282 = -0.229982 + N435*0.707414 + N435*N527*1.36263 - N435^2*0.40058 + N527^2*0.212145

N435 = 0.132348 + "Material Life Time"*0.271845 - "Material Life Time"*N499*0.0330638 + "Material Life Time"^2*0.0711447 + N499^2*0.80862

N499 = 0.566289 + Speed*0.410623 - Speed*"Maintenance Cost"*0.0933702 - "Maintenance Cost"*0.320342

N460 = 0.138407 - Corrodibility*N506*0.409389 + N506*0.669649 + N506^2*0.471662

N506 = 0.715386 + "Material Life Time"*"Maintenance Cost"*0.171258 + "Material Life Time"^2*0.21299 - "Maintenance Cost"*0.47548

N407 = -0.17629 + N466*0.671252 + N466*N470*0.62366 - N466^2*0.249816 + N470*0.467292 - N470^2*0.0991647

N470 = 0.0811291 + N510*0.122909 + N510*N516*1.32003 - N510^2*0.0846212

N516 = 0.503242 + Speed*0.393316 - "Operational Cost"*0.348945 + "Operational Cost"^2*0.176458

N510 = 0.981214 - "Operational Cost"*0.413576 + "Operational Cost"*"Maintenance Cost"*0.28439 + "Operational Cost"^2*0.0766547 - "Maintenance Cost"*0.54748

N466 = -0.565496 + N505*2.64513 - N505^2*1.34601 - N533*0.800566 + N533^2*1.58189

N351 = 0.0781319 + N429*0.520345 + N429*N494*1.03435 - N429^2*0.241478 - N494*0.418804 + N494^2*0.471358

N494 = -0.411628 + N515^2*0.866356 + N533*1.155

N515 = 0.909102 - "Install Cost(IC)"*0.232668 + "Install Cost(IC)"*"Maintenance Cost"*0.229891 - "Maintenance Cost"*0.508673

N429 = -0.998151 + N513*1.60441 - N513*N519*1.02177 + N519*1.75812 - N519^2*0.124941

N513 = 0.472428 + Speed*0.553123 - Speed*Corrodibility*0.0577999 - Speed^2*0.130839 - Corrodibility*0.260417 + Corrodibility^2*0.0672042

N384 = -0.417391 + N483*0.73122 + N483*N508*0.418399 + N508*0.720796

N508 = 0.124186 + N519*1.03015 + N519^2*0.00399616 - N533*1.60525 + N533^2*2.25891

N533 = 0.658713 - Corrodibility*0.245647 + Corrodibility*"Shaft Efficiency"*0.0469687 + "Shaft Efficiency"*0.0906693

N519 = 0.471285 + "Material Life Time"*0.572098 - "Material Life Time"*"Operational Cost"*0.188826 - "Material Life Time"^2*0.139348 - "Operational Cost"*0.33537 + "Operational Cost"^2*0.267681

N483 = -0.220463 - "Maintenance Cost"*N523*0.658071 + "Maintenance Cost"^2*0.0239713 + N523*2.1284 - N523^2*0.712782

N523 = 0.350014 + Speed*0.400877 - Speed*"Shaft Efficiency"*0.0173903 + "Shaft Efficiency"^2*0.122594

N241 = -0.275016 + N492*1.29554 - N492*N297*0.621367 - N492^2*0.747809 + N297*0.689907 + N297^2*0.524594

N297 = -0.169308 + N439*0.576362 + N439*N527*1.35929 - N439^2*0.296482 - N527*0.0789055 + N527^2*0.280978

N527 = 0.465723 - N534*1.57677 + N534*N538*3.71633 + N534^2*0.315318 - N538^2*1.01096

N538 = 0.654697 + "Shaft Efficiency"*0.121531 + "Shaft Efficiency"*"Operational Cost"*0.0201839 - "Operational Cost"*0.380354 + "Operational Cost"^2*0.191503

N534 = 0.80277 - Corrodibility*0.358846 + Corrodibility^2*0.128731 - "Install Cost(IC)"*0.149226

N439 = 0.207609 - "Maintenance Cost"*0.438571 - "Maintenance Cost"*N505*0.114627 + "Maintenance Cost"^2*0.156681 + N505*0.991666

N505 = 0.296459 + Speed*"Material Life Time"*0.0388992 + Speed^2*0.334978 + "Material Life Time"*0.322249

N492 = -1.0264 + N521*1.72242 - N521*N524*1.20113 + N524*1.7171

N524 = 0.292081 + "Shaft Efficiency"*0.23953 - "Shaft Efficiency"*"Material Life Time"*0.182787 + "Material Life Time"*0.442443

N521 = 0.398857 + Speed*0.492988 - Speed*"Install Cost(IC)"*0.207845 - "Install Cost(IC)"^2*0.0231703

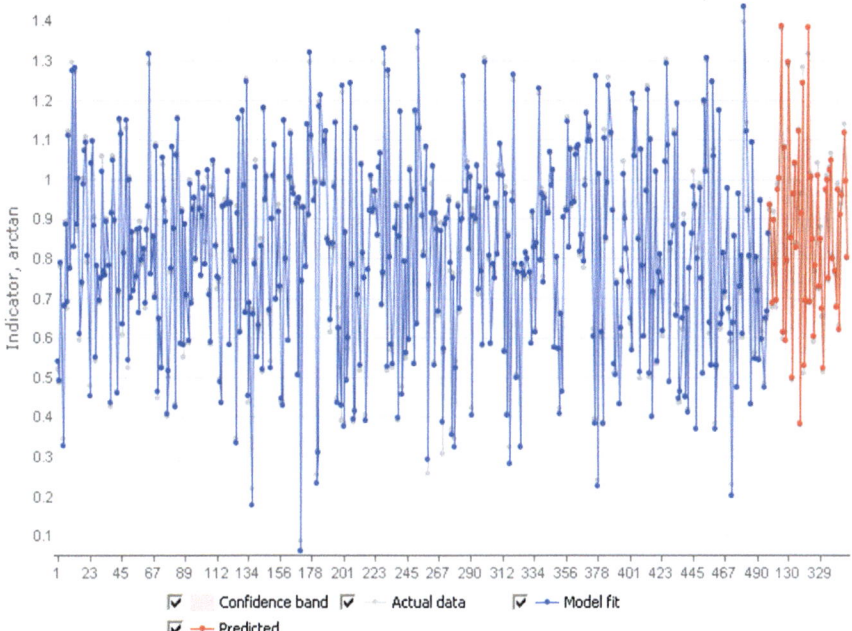

Figure showing comparison between Observed and Predicted Output for Model No. 7ANGAO1

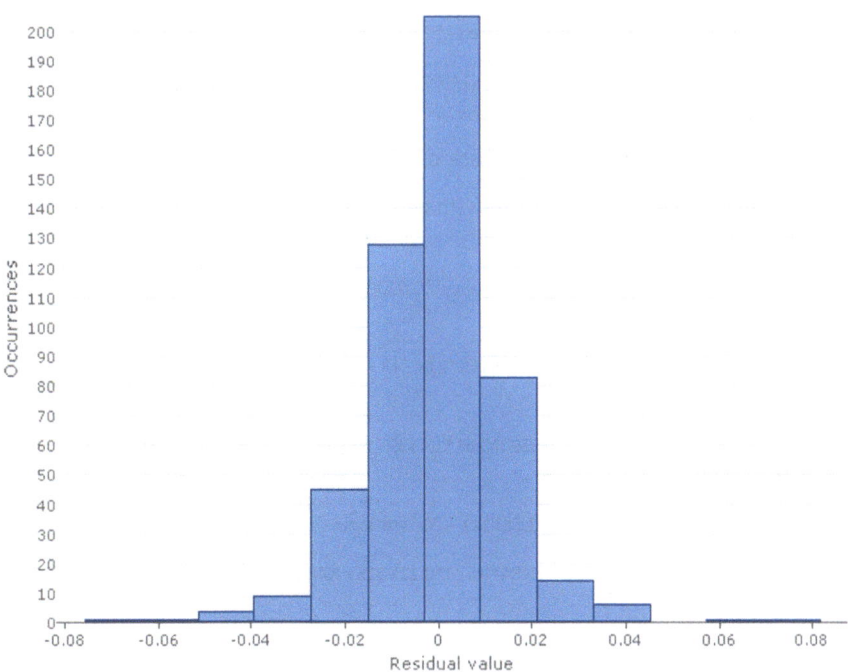

Fig. 5.4 Figure showing distribution of residuals from observed and predicted output for model No. 7ANGAO1

Model Equation #5.2. for Model No. 7ANGAO1

$Y1 = -0.00110758 - \text{"Install Cost(IC)"}*N2*0.00908735 + N2*1.00923 - N2^2*0.00388641$

$N2 = -0.00135302 - N44*N3*0.288705 + N3*1.00492 + N3^2*0.284921$

$N3 = -0.000943231 - N295*N4*0.0886409 + N295^2*0.0917599 + N4*0.998369$

$N4 = 0.120013 - N547*0.532135 - N547*N5*0.334887 + N547^2*0.499079 + N5*1.22272 + N5^2*0.0307026$

$N5 = 0.0100341 - N408*0.184101 - N408*N6*1.14171 + N408^2*0.673201 + N6*1.15233 + N6^2*0.485724$

$N6 = -0.00681153 + N419*0.0427389 + N7*0.965616$

$N7 = -0.0214015 + \text{"Install Cost(IC)"}*0.0702889 - \text{"Install Cost(IC)"}*N9*0.0481914 - \text{"Install Cost(IC)"}^2*0.0381179 + N9*1.02226$

$N9 = 0.00548297 - N171*0.182928 + N171*N11*0.00562036 + N11*1.17121$

$N11 = -0.00406557 - N237*0.547921 - N237*N14*4.49918 + N237^2*2.46731 + N14*1.55987 + N14^2*2.02099$

$N14 = -0.00183078 + N22*0.486547 + N22^2*0.0928494 + N35*0.517355 - N35^2*0.0947201$

$N35 = 0.00800251 - N113*0.873157 - N113*N56*1.47598 + N113^2*1.48983 + N56*1.85089$

$N56 = -0.00588551 + N79*0.0927496 + N79*N97*0.500678 + N97*0.920929 - N97^2*0.507658$

$N97 = 0.00763767 + N300*0.31031 + N300*N140*2.39775 - N300^2*1.20687 + N140*0.663777 - N140^2*1.16431$

$N300 = -0.914668 + N478*1.15598 - N478*N503*0.262773 + N503*1.26196 - N503^2*0.0961623$

$N478 = -1.13003 + N486*3.17419 - N486*N543*0.379434 - N486^2*1.13447 - N543*0.396089 + N543^2*1.08191$

$N543 = 0.846261 + \text{Corrodibility}*\text{"Shaft Efficiency"}*0.0200339 - \text{Corrodibility}^2*0.229345 + \text{"Shaft Efficiency"}^2*0.113482$

$N79 = 0.0386394 - N422*0.58545 - N422*N145*2.04343 + N422^2*1.26668 + N145*1.49911 + N145^2*0.813072$

$N145 = -0.71696 + N539*2.55969 + N539*N261*1.4339 - N539^2*2.26338 + N261*0.210204 - N261^2*0.231639$

$N539 = 1.00934 - \text{"Operational Cost"}*0.581141 + \text{"Operational Cost"}^2*0.279948$

$N422 = -0.825391 + N484*2.10887 + N484*N524*0.119015 - N484^2*0.760354 + N524*0.122329 + N524^2*0.404684$

N113 = 0.150074 - N484*0.746088 - N484*N154*0.821071 + N484^2*0.821466 + N154*1.39114 + N154^2*0.192819

N154 = -0.368173 + "Operational Cost"*0.597031 - "Operational Cost"*N261*0.443754 - "Operational Cost"^2*0.233529 + N261*1.59094 - N261^2*0.226499

N261 = -0.0955126 + N428*0.689483 + N428*N526*0.505384 - N428^2*0.0817124 - N526*0.339244 + N526^2*0.512256

N428 = -1.64267 + N484*2.79279 - N484*N535*0.425288 - N484^2*0.88488 + N535*1.28061 + N535^2*0.0586624

N22 = -0.106355 + "Operational Cost"*0.266287 - "Operational Cost"*N59*0.122959 - "Operational Cost"^2*0.167965 + N59*1.13365 - N59^2*0.0446589

N59 = 0.0274799 + Speed*0.177531 - Speed*N83*0.560477 + Speed^2*0.258729 + N83*0.801929 + N83^2*0.299907

N83 = -0.0112064 + N161*0.542139 + N161*N197*1.17274 - N161^2*0.600928 + N197*0.483211 - N197^2*0.582725

N161 = 0.211683 + N528*N267*1.18187 - N528^2*0.533142 + N267*0.438273 - N267^2*0.26526

N237 = -0.0162755 - N283*0.399146 + N283*N289*1.39752 + N289*1.4587 - N289^2*1.43971

N289 = -0.677175 + N427*1.09955 - N427^2*0.0964124 + N522*0.815411

N283 = -0.69274 + N452*0.970866 + N452*N522*0.145309 - N452^2*0.0703698 + N522*0.82151

N522 = 1.6007 + N536*N545*1.24017 - N545*4.17543 + N545^2*2.67532

N536 = 0.900356 - "Shaft Efficiency"*"Maintenance Cost"*0.139146 + "Shaft Efficiency"^2*0.218029 - "Maintenance Cost"*0.296177 + "Maintenance Cost"^2*0.0853047

N452 = -0.30367 - "Operational Cost"*0.551623 + "Operational Cost"^2*0.248223 + N484*2.2834 - N484^2*0.788711

N484 = 0.383838 + Speed*0.62704 + "Material Life Time"*0.202355 - "Material Life Time"^2*0.00583957

N171 = 0.483389 - N506*1.06803 + N506*N265*1.16048 + N506^2*0.0898146 + N265*0.921283 - N265^2*0.560482

N265 = -1.50983 + N441*2.44889 - N441*N530*0.414715 - N441^2*0.649792 + N530*1.3078

N530 = 3.46433 - N545*6.52937 + N545*N547*3.91767 + N545^2*2.77732 - N547*2.19897

N506 = 1.78816 + N524*N545*1.19616 - N545*4.57567 + N545^2*2.92536

N419 = -0.380521 + N485*0.916253 + N485^2*0.0585343 + N528^2*0.601766

N528 = 1.15256 - Corrodibility*0.293813 + Corrodibility*"Operational Cost"*0.113259 - "Operational Cost"*0.640975 + "Operational Cost"^2*0.285371

N408 = -1.36511 + N441*3.20393 - N441*N546*1.27197 - N441^2*0.673029 + N546^2*1.33117

N546 = 4.28212 + N547*4.13026 - N547^2*1.97678 - N550*13.4297 + N550^2*8.1638

N441 = 0.278252 + Speed*N524*0.738409 + N524^2*0.32532

N295 = -0.389738 + N454*1.0282 - N454^2*0.0285042 + N523^2*0.570142

N523 = 0.557566 - "Maintenance Cost"*0.360044 + "Maintenance Cost"^2*0.0910585 + N537^2*0.607736

N537 = 0.749095 - Corrodibility*0.129182 - Corrodibility*"Material Life Time"*0.209737 + "Material Life Time"*0.472639 - "Material Life Time"^2*0.182822

N454 = -1.56633 + N486*2.6602 - N486*N535*0.288923 - N486^2*0.865594 + N535*1.23831

N44 = 0.0212164 - N486*0.2954 - N486*N81*0.957094 + N486^2*0.647684 + N81*1.2407 + N81^2*0.331026

N81 = -0.00547544 + N140*0.723441 + N140*N197*2.28661 - N140^2*1.23735 + N197*0.293356 - N197^2*1.05718

N197 = -0.0162949 + N269*0.48664 + N269*N276*0.666574 - N269^2*0.288337 + N276*0.540275 - N276^2*0.384119

N276 = -0.780996 + N475*1.11951 - N475*N503*0.135423 - N475^2*0.0442001 + N503*0.987638

N503 = 1.16724 - N535*0.915777 + N535*N540*2.44439 - N540*2.17047 + N540^2*0.804043

N540 = 0.925907 + "Material Life Time"*"Maintenance Cost"*0.210653 + "Material Life Time"^2*0.0615854 - "Maintenance Cost"*0.370243

N535 = 0.944544 + "Shaft Efficiency"*0.137638 + "Shaft Efficiency"*"Operational Cost"*0.0409425 - "Operational Cost"*0.590207 + "Operational Cost"^2*0.253993

N475 = 1.3318 + N485*0.594948 + N485*N545*0.487625 - N545*4.0622 + N545^2*2.96084

N485 = 0.283936 + Speed*1.20051 - Speed*"Shaft Efficiency"*0.216886 - Speed^2*0.447801 + "Shaft Efficiency"*0.217079 + "Shaft Efficiency"^2*0.038955

N269 = -2.62643 + N411*2.08743 - N411*N548*0.894158 - N411^2*0.219357 + N548*4.4219 - N548^2*1.65788

N548 = -4.60438 + N549*2.70985 - N549*N550*2.04238 + N550*8.33408 - N550^2*3.34299

N550 = 0.773062 + "Shaft Efficiency"^2*0.129558

N549 = 0.714376 + "Material Life Time"*0.319949 - "Material Life Time"*"Install Cost(IC)"*0.252592 - "Install Cost(IC)"^2*0.000617047

N411 = 0.0302258 + N482*N524*1.17941

N524 = 1.19937 - "Operational Cost"*0.607703 + "Operational Cost"*"Maintenance Cost"*0.203366 + "Operational Cost"^2*0.1852 - "Maintenance Cost"*0.386116

N482 = 0.513192 + Speed*1.13191 - Speed^2*0.489515 - Corrodibility*0.228782 - Corrodibility^2*0.0116342

N140 = -0.379573 + "Operational Cost"*0.582253 - "Operational Cost"*N267*0.510858 - "Operational Cost"^2*0.183296 + N267*1.63348 - N267^2*0.235407

N267 = -0.0945882 + N427*0.829537 + N427*N526*0.402975 - N427^2*0.12131 - N526*0.456876 + N526^2*0.626118

N526 = 1.46361 + N542*N545*1.27924 - N545*3.86809 + N545^2*2.46764

N545 = 0.908966 - Corrodibility*"Install Cost(IC)"*0.258424 - Corrodibility^2*0.101449

N542 = 0.950006 - "Maintenance Cost"*0.266722

N427 = -3.23146 + N476*2.90889 - N476*N547*1.41906 - N476^2*0.453657 + N547*5.02473 - N547^2*1.72978

N547 = 0.593665 + "Shaft Efficiency"*0.229624 - "Shaft Efficiency"*"Material Life Time"*0.158084 + "Material Life Time"*0.284113

N476 = 0.669315 + Speed*0.641702 - Speed*"Operational Cost"*0.0240279 - "Operational Cost"*0.522325 + "Operational Cost"^2*0.224515

N486 = 0.488565 + Speed*0.717358 - Speed*"Install Cost(IC)"*0.18665

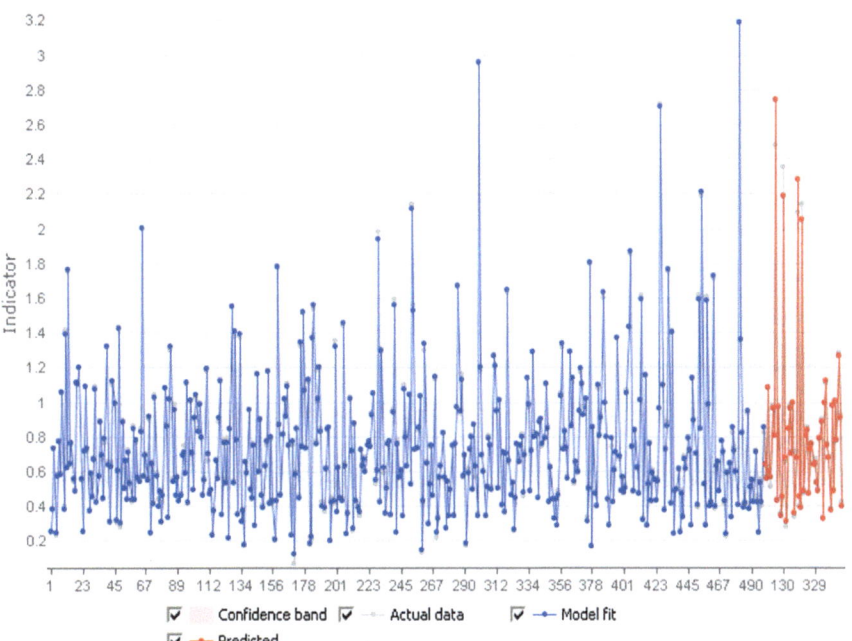

Figure showing comparison between Observed and Predicted Output for Model No. 7AHGN1

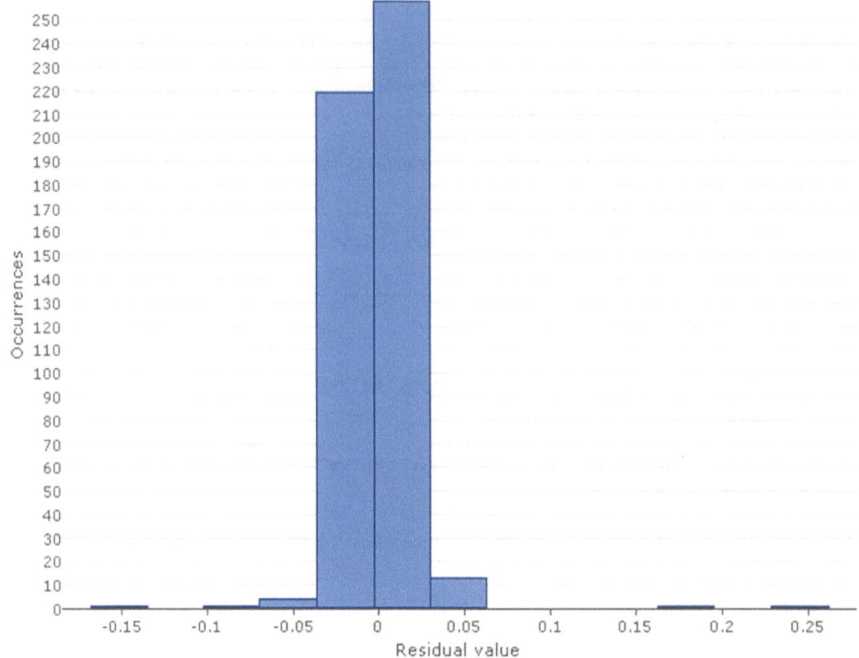

Figure showing distribution of residuals from Observed and Predicted Output for Model No. 7AHGN1

Model Equation for Model No. 7AHGN1:

Y1 = -0.0120557 + N473*0.0304614 - N473*N2*0.0289041 + N2*1.00019 + N2^2*0.0105556

N2 = -0.00241215 + "Operational Cost"*0.0346218 - "Operational Cost"^2*0.0373317 + N3*0.994156 + N3^2*0.00273048

N3 = 0.017051 - N502*0.0746048 - N502*N4*0.0265044 + N502^2*0.0648806 + N4*1.021

N4 = 0.0641277 - N507*0.178393 - N507*N5*0.00999574 + N507^2*0.118021 + N5*1.00774

N5 = 0.00929928 - N400*0.0483859 + N400^2*0.0219095 + N6*1.02497 - N6^2*0.00920677

N6 = -0.00584021 + N195*0.132775 + N195*N7*0.427545 - N195^2*0.286505 + N7*0.881923 - N7^2*0.147424

N7 = 0.00352067 - N276*0.120607 - N276*N8*0.149048 + N276^2*0.103748 + N8*1.11343 + N8^2*0.0472888

N8 = 0.0106289 - "Install Cost(IC)"*0.0684432 + "Install Cost(IC)"^2*0.0686668 + N9*1.00136 - N9^2*0.000550789

N9 = 0.00760009 - N26*1.23082 + N26*N10*0.883271 + N10*2.21341 - N10^2*0.875572

N10 = -0.000700408 + N12*N15*10.2406 - N12^2*4.83627 + N15*1.00221 - N15^2*5.40404

N15 = -0.0534073 + N461*0.181605 + N461*N18*0.0450515 - N461^2*0.116153 + N18*0.951042

N18 = -0.010128 + N478*0.0101855 - N478*N26*0.102651 + N478^2*0.0743612 + N26*0.995137 + N26^2*0.0331833

N461 = -0.511878 + "Maintenance Cost"*0.593318 - "Maintenance Cost"*N507*1.81303 + N507*2.45379 - N507^2*0.317763

N12 = -0.0253111 + "Shaft Efficiency"*0.13017 - "Shaft Efficiency"^2*0.118369 + N20*0.999975

N20 = -0.00588864 + N448*0.00872738 - N448*N25*0.0797308 + N448^2*0.0607856 + N25*0.987657 + N25^2*0.0286683

N25 = -0.0073384 + N46*0.640172 - N46*N63*6.78483 + N46^2*3.37676 + N63*0.375562 + N63^2*3.39753

N63 = -0.0479758 + N404*0.272988 - N404^2*0.0937539 + N75*0.832527 + N75^2*0.0450532

N404 = 0.475146 - N494*0.459863 + N494*N506*1.60321 + N494^2*0.167186 - N506*0.839516 + N506^2*0.412121

N494 = 1.02603 - "Material Life Time"*"Maintenance Cost"*0.0172182 + "Material Life Time"^2*0.479401 - "Maintenance Cost"*1.29527 + "Maintenance Cost"^2*0.598628

N46 = -0.0048413 + N121*1.0083 - N121*N122*0.140914 - N121^2*0.15097 + N122^2*0.289809

N122 = -0.158467 + N486*0.332322 - N486*N188*0.513693 + N486^2*0.0404311 + N188*1.1395 + N188^2*0.14294

N121 = -0.0170149 + N280*0.527731 + N280*N137*0.240924 - N280^2*0.256032 + N137*0.510265

N137 = -0.00265735 + N252*1.10248 + N252*N281*0.410453 - N252^2*0.399992 - N281*0.102296

N281 = 0.176987 - N416*0.425888 + N416*N470*1.18741 + N416^2*0.153956 + N470*0.366979 - N470^2*0.263639

N470 = -0.296522 + N503*0.765234 + N503*N513*0.314452 + N513^2*0.511422

N280 = -0.0109872 - N433*0.297818 + N433*N486*1.2773 + N433^2*0.134564 + N486*0.5303 - N486^2*0.349689

N433 = 0.306251 - N496*0.90821 + N496*N507*2.37068 + N496^2*0.116518 - N507^2*0.448652

N496 = 1.08378 + "Shaft Efficiency"*0.112891 - "Shaft Efficiency"*"Maintenance Cost"*0.324769 + "Shaft Efficiency"^2*0.287756 - "Maintenance Cost"*1.27964 + "Maintenance Cost"^2*0.696757

N448 = -0.75827 + N503*1.44725 + N503*N506*0.471829 - N503^2*0.524565 + N506*0.644693

N26 = -0.00713927 - N40*0.169337 - N40*N54*0.686394 + N40^2*0.679418 + N54*1.1854

N54 = -1.47067e-05 + N105*0.556631 + N105^2*0.00298857 + N131*0.440518

N131 = -0.117449 + N418*0.364614 - N418*N165*0.549436 + N418^2*0.0453236 + N165*0.989014 + N165^2*0.259764

N165 = 0.0262658 + N252*0.623735 + N252*N271*1.25141 - N252^2*0.609785 + N271*0.287367 - N271^2*0.571837

N271 = 0.369424 - N446*0.83559 + N446*N478*1.43676 + N446^2*0.3939 - N478^2*0.0828929

N478 = -0.00466442 + N506*N515*1.33697 + N515^2*0.0250777

N506 = 0.686199 + Speed*0.433449 + Speed^2*0.193886 - "Operational Cost"*0.776148 + "Operational Cost"^2*0.42399

N446 = -0.0440221 - Corrodibility*N500*0.689526 + N500*1.3986

N252 = 0.150456 - N431*0.494886 + N431*N492*1.34234 + N431^2*0.23371 + N492*0.241141 - N492^2*0.183928

N418 = 0.053439 + N482*N515*0.845071 + N482^2*0.215946 - N515*0.17787 + N515^2*0.383256

N105 = 0.00439897 + N145*0.983467 - N145*N195*1.06187 + N145^2*0.402046 + N195^2*0.666964

N40 = -0.00365547 + N75*0.463682 - N75*N120*2.22517 + N75^2*1.14572 + N120*0.540254 + N120^2*1.07653

N120 = -0.174146 + N492*0.373898 - N492*N188*0.517026 + N492^2*0.0200328 + N188*1.1385 + N188^2*0.14198

N188 = -0.0259426 + N292*0.269584 - N292*N314*0.29561 + N292^2*0.27038 + N314*0.758716 + N314^2*0.0255693

N314 = 0.417335 - N391*0.677091 + N391*N517*2.03712 + N391^2*0.110072 - N517*0.634128 + N517^2*0.194248

N517 = 0.259063 - N520*0.181957 + N520*N522*2.1619 - N520^2*0.351571 - N522^2*0.67876

N522 = 0.652636 + "Shaft Efficiency"*0.37908 - "Shaft Efficiency"*"Install Cost(IC)"*0.405334

N520 = 0.899327 + "Shaft Efficiency"*0.104482 - "Shaft Efficiency"*"Operational Cost"*0.0111528 + "Shaft Efficiency"^2*0.112837 - "Operational Cost"*0.791356 + "Operational Cost"^2*0.439135

N391 = 0.304025 - N482*0.464652 + N482*N503*1.62032 + N482^2*0.171689 - N503*0.414862 + N503^2*0.158824

N503 = 0.59634 - Corrodibility*0.334409 - Corrodibility*"Material Life Time"*0.325358 + Corrodibility^2*0.0865992 + "Material Life Time"*0.689842

N482 = 0.755735 + Speed*0.825057 - Speed*"Maintenance Cost"*0.485474 - "Maintenance Cost"*1.0701 + "Maintenance Cost"^2*0.645644

N292 = -0.0604639 - N422*0.359355 + N422*N486*0.992481 + N422^2*0.24827 + N486*0.853136 - N486^2*0.451994

N492 = -0.271351 + N513*N516*0.38811 + N513^2*0.472929 + N516*0.707024

N75 = 0.00979633 + N145*0.970536 - N145*N171*0.909519 + N145^2*0.331092 + N171^2*0.591988

N171 = 0.26481 - N422*0.663001 + N422*N269*0.666454 + N422^2*0.102427 + N269*0.973485 - N269^2*0.346925

N422 = 0.941781 - N495*1.00989 + N495*N512*2.72416 - N512*1.55881 + N512^2*0.364323

N512 = 1.16212 - Corrodibility*0.754877 + Corrodibility*"Install Cost(IC)"*0.179113 + Corrodibility^2*0.228305 - "Install Cost(IC)"*0.270614 - "Install Cost(IC)"^2*0.110196

N145 = -0.0111983 + N244*1.05856 - N244^2*0.214949 - N266*0.0384987 + N266^2*0.21065

N266 = 0.0780431 - N425*0.389414 + N425*N468*1.17031 + N425^2*0.145472 + N468*0.582958 - N468^2*0.378001

N468 = 0.00341114 + N501*N515*0.866419 + N501^2*0.213651 + N515^2*0.234926

N501 = 0.554294 + Speed*0.808848 - Speed*Corrodibility*0.261383 - Speed^2*0.0548814 - Corrodibility*0.373516 + Corrodibility^2*0.0924729

N425 = 1.19349 - N469*1.79142 + N469*N521*3.29291 + N469^2*0.210236 - N521*1.431

N521 = 1.14976 - "Install Cost(IC)"*0.366407 + "Install Cost(IC)"*"Operational Cost"*0.21959 - "Operational Cost"*0.836534 + "Operational Cost"^2*0.394671

N469 = 1.67831 - Corrodibility*0.858989 + Corrodibility*"Maintenance Cost"*0.791275 - "Maintenance Cost"*1.95486 + "Maintenance Cost"^2*0.81486

N244 = 0.135746 - N431*0.564887 + N431*N486*1.36539 + N431^2*0.254346 + N486*0.369398 - N486^2*0.27407

N486 = -0.00594545 + N509*N515*0.581414 + N509^2*0.351408 + N515^2*0.384251

N431 = 0.391388 - N495*0.57553 + N495*N514*2.17944 - N514*0.621263 + N514^2*0.0805678

N276 = 0.389853 - N419*0.381929 + N419*N502*1.63447 + N419^2*0.108354 - N502*0.790189 + N502^2*0.432154

N419 = 0.195736 - N495*0.505977 + N495*N509*0.989384 + N495^2*0.405787 + N509*0.16957

N509 = 0.418087 + Speed*0.84043 - Speed*"Install Cost(IC)"*0.487218 + Speed^2*0.0281727 - "Install Cost(IC)"^2*0.00624912

N195 = 0.17865 - N416*0.427809 + N416*N269*0.638837 + N269*0.951642 - N269^2*0.330638

N269 = 0.0029813 - N427*0.0993669 + N427*N497*1.45309 + N497*0.172332 - N497^2*0.195382

N497 = 0.53973 - N514*1.58817 + N514*N516*1.75847 + N514^2*0.917339 - N516^2*0.177904

N516 = 0.457308 + "Material Life Time"*0.790908 - "Material Life Time"*"Install Cost(IC)"*0.50163

N427 = 0.107807 + N495*N513*1.30927 - N513*0.253925 + N513^2*0.175407

N416 = 0.511651 - N500*1.51717 + N500*N507*2.78183 + N500^2*0.309254 - N507^2*0.613822

N500 = 1.49811 - "Install Cost(IC)"*0.555782 + "Install Cost(IC)"*"Maintenance Cost"*0.624379 - "Maintenance Cost"*1.75638 + "Maintenance Cost"^2*0.727492

N400 = 0.15945 - N495*0.386843 + N495*N499*1.21559 + N495^2*0.236797 + N499*0.0712089

N499 = 0.373743 + Speed*"Material Life Time"*0.352429 + Speed^2*0.416457 + "Material Life Time"^2*0.340279

N495 = 1.62328 - "Operational Cost"*0.808772 + "Operational Cost"*"Maintenance Cost"*0.865467 - "Maintenance Cost"*1.79704 + "Maintenance Cost"^2*0.629595

N507 = 1.35684 - Corrodibility*0.914482 + Corrodibility*"Operational Cost"*0.500845 + Corrodibility^2*0.241279 - "Operational Cost"*1.08635 + "Operational Cost"^2*0.502015

N502 = -0.252247 + N514*N515*1.16316 + N515*0.981731 - N515^2*0.649661

N515 = 0.282841 + "Shaft Efficiency"*0.331865 - "Shaft Efficiency"*"Material Life Time"*0.202348 + "Material Life Time"*0.658865

N514 = 0.868187 - Corrodibility*0.499596 - Corrodibility*"Shaft Efficiency"*0.0522118 + Corrodibility^2*0.108036 + "Shaft Efficiency"*0.162026 + "Shaft Efficiency"^2*0.0376469

N473 = 0.214056 - "Maintenance Cost"*0.77876 - "Maintenance Cost"*N513*0.769912 + "Maintenance Cost"^2*0.656134 + N513*1.32911

N513 = 0.335121 + Speed*0.674662 - Speed*"Shaft Efficiency"*0.084573 + "Shaft Efficiency"^2*0.224482

5.1 Scientific Implications

The present investigation provides an opportunity to onsite engineers for ready-made analysis of turbine performance.

The process can reduce cost of monitoring and maintenance as real time monitoring and early warning both can be possible from the implementation of the indicator in real life situations.

The replacement period and the payback period of turbine is a complex procedure for estimation but it is mandatory in many captive power plants.

The indicator and its platform independent model can easily provide an opportunity to predict both payback and replacement period.

Other than the benefits previously discussed the suitability of turbines for utilization in the power plant and priority parameters which changes the turbine efficiency can be identified with this method.

5.2 Assumption and Limitation

The model methodology although have a lot of strength like its objectivity and cognitivity but still as is the case for all the numerical models this model have some limitations:

The number and type of criteria and alternatives if changed with in the method, then the same model can yield erroneous results.

The type of MCDM used for this purpose can also make the study questionable if the method of MCDM is changed.

The number of hidden layers or types of activation function selected will also modify the results of turbine vulnerability.

References

Ciubotariu CR, Secosan E, Marginean G, Frunzaverde D, Campian V (2016) Experimental study regarding the cavitation and corrosion resistance of Stellite 6 and self-fluxing remelted coatings

Das HS, Yatim AHM, Tan CE, Lau KY (2016) Proposition of a PV/tidal powered micro-hydro and diesel hybrid system: a southern Bangladesh focus. Renew Sustain Energ Rev 53 (2016):1137–1148

Gao B, Zhang N, Li Z, Ni D, Yang M (2016) Influence of the blade trailing edge profile on the performance and unsteady pressure pulsations in a low specific speed centrifugal pump. J Fluids Eng 138(5):051106

Marengo H, Arreguin FI, Aldama AA, Morales V (2013) Case study: risk analysis by overtopping of diversion works during dam construction: the La Yesca hydroelectric project, Mexico. Struct Saf 42:26–34

Martins GL, Braga SL, Ferreira SB (2016) Design optimization of partial admission axial turbine for ORC service. Appl Therm Eng 96(2016):18–25

Phillips WD, Staniewski JWG (2016) The origin, measurement and control of fine particles in non-aqueous hydraulic fluids and their effect on fluid and system performance. Lubr Sci 28 (1):43–64

Yang W, Norrlund P, Saarinen L, Yang J, Guo W, Z Wei (2016) Wear and tear on hydro power turbines—influence from primary frequency control. Renew Energy 87:88–95

Zawahri NA (2009) India, Pakistan and cooperation along the Indus River system. Water Policy 11 (1):1–20

Chapter 6
Conclusion

Abstract The study was concluded with the observations that the method dependency of the model may be reduced to produce uniform results. As a future research scope the same model can be applied to design a early warning or real time monitoring system which can greatly reduce the system uncertainty.

Keywords Turbine vulnerability · Indicator · GMDH

The present investigation is an attempt to devise a system which can represent climatic vulnerability on the performance efficiency of hydro-turbines.

In this aspect the present study utilized three MCDM methods and one cognitive method namely AHP, ANP, MACBETH and GMDH.

According to the results the Specific Speed of turbines, Corrosiveness of the turbine blades operational and maintenance cost along with installation cost and material lifetime was found to be the six most important variables in influencing the performance of the runner. All the three methods were unanimous in this regard although the ranking is different for different methods. The specific speed however was found to be the most important variables as determined by two out of three MCDM methods used for identification of priority parameters and priority values with respect to the study objective.

The priority value and magnitude of the priority parameters were used to develop the indicator where the variables which are directly proportional to turbine performance was placed in the numerator and rest of the variables which ditoreates or increase the cost of operation was kept in the denominator so that the indicator becomes directly proportional to the turbine performance.

In this regard a neural network model was also developed to map the input priority parameters with the output indicator. In total twenty four different models were developed where eight models each were developed with the priority values per MCDM method. In total twenty four models were developed with seven and three most important parameter as input. The output data was also transformed by Arc Tangent function for twelve different models and for the rest of the models data of output was kept unchanged.

U. Roy and M. Majumder, *Impact of Climate Change on Small Scale Hydro-turbine Selections*, SpringerBriefs in Energy, DOI 10.1007/978-981-287-239-5_6

The mean absolute error and correlation coefficient was used to identify the three models which have the highest performance and within this three models the best model was selected with the help of Root Mean Square Error (RMSE), Covariance (Cov), Standard Deviation (SD) and Mean Relative Error (MRE).

It was found that the model with all the seven priority parameters as input, data of output transformed by Arc Tangent function, trained with GMDH algorithm and priority values of the parameters retrieved with AHP MCDM have the highest equivalent performance index which was derived by a function which is inversely proportional to RMSE, Cov, SD and MRE. The function gives more weightage to the metrics of testing compared to the metrics of training.

The sensitivity analysis of the model also revealed that the most sensitive input was also the most important parameter as estimated by the MCDM method from which the priority value of the model output was collected. The accuracy of the better model among the twenty four models used for the study was found to be above 98 %.

The model was tested in six different hydropower plants located in different places of the World. The impact of climate change on the turbine performance was also analyzed for IPCC A2 and B2 scenario.

The data for the climatic parameters like temperature and precipitation were derived from CMIP5 model and the data for the input parameters of the present model was generated with the help of a new model developed with the GMDH algorithm. The model tries to establish a relationship between the climatic parameters and change in population with corrodibility of water of the location and operational and maintenance expenditure of the system using the turbine.

The impact of climate change on the turbines located in six different selected study areas were analyzed and from the results it was found that in most of the places most and least vulnerable period was in the first three decades of A2 and last three decades of B2 respectively.

6.1 Strength and Weakness of the Model

The model developed for the present study yielded two important benefits like:

(1) The model methodology helped to estimate the most important variable for analysis of turbine performances. The application of the MCDM method imbibed a logical process of identification of the most influential variable in determination of turbine performance.

(2) The next benefit was the model developed for the study creates a single point indicator for representation of turbine performance. As the model has utilized the advantage of ANN methodology the indicator become cognitive in nature.

However the developed methodology has some specific weaknesses like:

(1) The dependency of the model on the result from MCDM and GMDH method may often reduce the exclusivity of the model. Although in this investigation three different MCDM was compared and best result from these three methods were only selected.
(2) Many other factors can affect the efficiency of turbines as certain number of literatures is only surveyed. The number and type of criteria and alternative may also change the model output. In the present study a thorough literature survey was performed and the parameters were selected by its citation frequency.
(3) In case of ANN models, although high level of accuracy was achieved by the models but reliability of dataset will always influence the accuracy of the model as is the case for the models developed by neural networks.

6.2   Future-scope

As a future pathway of research in this field the researchers can think about solving the problems of method dependency and survey of literatures for deriving the important parameters by the utilization of some more objective methods.

The problem with ANN models can be also be solved by introducing some stoichiometric methodology for generation of the training dataset.

The model developed in the present study can be used in an Early Warning or Real Time Monitoring systems so that uncertainties can be identified real time and proper mitigation measures can be adopted to save in maintenance expenditures and maintain efficiency of the power plants.